Oliver Wendell Holmes

Border Lines of Knowledge in Some Provinces of Medical Science

An Introductory Lecture, Delivered Before the Medical Class of Harvard

University

Oliver Wendell Holmes

Border Lines of Knowledge in Some Provinces of Medical Science
An Introductory Lecture, Delivered Before the Medical Class of Harvard University

ISBN/EAN: 9783337179854

Printed in Europe, USA, Canada, Australia, Japan

Cover: Foto ©berggeist007 / pixelio.de

More available books at **www.hansebooks.com**

BORDER LINES OF KNOWLEDGE

IN SOME PROVINCES OF

MĒDICAL SCIENCE.

AN INTRODUCTORY LECTURE,

DELIVERED BEFORE THE MEDICAL CLASS OF HARVARD UNIVERSITY,
NOVEMBER 6TH, 1861.

BY OLIVER WENDELL HOLMES, M. D.,

PARKMAN PROFESSOR OF ANATOMY AND PHYSIOLOGY.

BOSTON:
TICKNOR AND FIELDS.
1862.

College Library, November 22, 1861.

PROFESSOR O. W. HOLMES: —

Dear Sir, — At a meeting of the Medical Class it was unanimously voted, that your Introductory Address be published. We therefore take much pleasure in requesting of you a copy for that purpose.

Truly your friends,

WM. J. RADFORD, *Chairman.*
W. K. FLETCHER, ⎫
W. H. MACDONALD, ⎬ *Committee.*
D. H. HAYDEN, ⎭

———

Boston, November 25, 1861.

GENTLEMEN: —

It pleases me to comply with the request you have addressed me, in accordance with the vote of the Medical Class, to furnish you a copy of my Introductory Lecture for publication.

Very truly your friend,

O. W. HOLMES.

Messrs. WM. J. RADFORD, *Chairman.*
W. K. FLETCHER, ⎫
W. H. MACDONALD, ⎬ *Committee.*
D. H. HAYDEN, ⎭

1

TO THE READER.

THIS Lecture appears as it would have been delivered had the time allowed been less strictly limited. Passages necessarily omitted have been restored, and points briefly touched have been more fully considered. A few notes have been added for the benefit of that limited class of students who care to track an author through the highways and by-ways of his reading. I owe my thanks to several of my professional brethren who have communicated with me on subjects with which they are familiar; especially to Dr. John Dean, for the opportunity of profiting by his unpublished labors, and to Dr. Hasket Derby, for information and references to recent authorities relating to the anatomy and physiology of the eye.

LECTURE.

THE entrance upon a new course of Lectures is always a period of interest to instructors and pupils. As the birth of a child to a parent, so is the advent of a new class to a teacher. As the light of the untried world to the infant, so is the dawning of the light resting over the unexplored realms of science to the student. In the name of the Faculty I welcome you, Gentlemen of the Medical Class, new-born babes of science, or lustier nurslings, to this morning of your medical life, and to the arms and the bosom of this ancient University. Fourteen years ago I stood in this place for the first time to address those who occupied these benches. As I recall these past seasons of our joint labors, I feel that they have been on the whole prosperous, and not undeserving of their prosperity. For it has been my privilege to be associated with a body of true and faithful workers; I cannot praise them freely to their faces, or I should be proud to discourse of the harmonious diligence and the noble spirit in which they have toiled together, not merely to teach their several branches, but to elevate the whole standard of teaching.

I may speak with less restraint of those gentlemen who have aided me in the most laborious part of my daily duties, the Demonstrators, to whom the successive classes have owed so much of their instruction. They rise before me, the dead and the living, in the midst of the most grateful recollections. The fair, manly face and stately figure of my friend, Dr. Samuel Parkman, himself fit for the highest offices of teaching, yet willing to be my faithful assistant in the time of need, come back to me with the long sigh of regret for his early loss to our earthly companionship. Every year I speak the eulogy of Dr. Ainsworth's patient toil as I show his elaborate preparations. When I take down my American Cyclopædia and borrow instruction from the learned articles of Dr. Kneeland, I cease to regret that his indefatigable and intelligent industry was turned into a broader channel. And what can I say too cordial of my long associated companion and friend, Dr. Hodges, whose admirable skill, working through the swiftest and surest fingers that ever held a scalpel among us, has delighted class after class, and filled our Museum with monuments which will convey his name to unborn generations?

This day belongs, however, not to myself and my recollections, but to all of us who teach and all of you who listen, whether experts in our specialties or aliens to their mysteries, or timid neophytes just entering the portals of the hall of science. Look in with me, then, while I attempt to throw some rays into its interior, which shall illuminate a few of its pillars and cornices, and show at the same time how many niches and alcoves remain in darkness.

SCIENCE is the topography of ignorance. From a few elevated points we triangulate vast spaces, enclosing infinite unknown details. We cast the lead, and draw up a little sand from abysses we shall never reach with our dredges.

The best part of our knowledge is that which teaches us where knowledge leaves off and ignorance begins. Nothing more clearly separates a vulgar from a superior mind, than the confusion in the first between the little that it truly knows, on the one hand, and what it half knows and what it thinks it knows, on the other.

That which is true of every subject is especially true of the branch of knowledge which deals with living beings. Their existence is a perpetual death and reanimation. Their identity is only an idea, for we put off our bodies many times during our lives, and dress in new suits of bones and muscles.*

> " Thou art not thyself;
> For thou exist'st on many a thousand grains
> That issue out of dust."

* " Occasio enim præceps est propter artis materiam, dico autem corpus, quod continue fluit et momento temporis transmutatur." — GALEN. Com. in Aphorism. Hippoc. l. 1.

If it is true that we understand ourselves but imperfectly in health, it is more signally manifest in disease, where natural actions imperfectly understood, disturbed in an obscure way by half-seen causes, are creeping and winding along in the dark toward their destined issue, sometimes using our remedies as safe stepping-stones, occasionally, it may be, stumbling over them as obstacles.

I propose in this lecture to show you some points of contact between our ignorance and our knowledge in several of the branches upon the study of which you are entering. I may teach you a very little directly, but I hope much more from the trains of thought I shall suggest. Do not expect too much ground to be covered in this rapid survey. Our task is only that of sending out a few pickets under the starry flag of science to the edge of that dark domain where the ensigns of the obstinate rebel, Ignorance, are flying undisputed. We are not making a reconnoissance in force, still less advancing with the main column. But here are a few roads along which we have to march together, and we wish to see clearly how far our lines extend, and where the enemy's outposts begin.

Before touching the branches of knowledge that deal with organization and vital functions, let us glance at that science which meets you at the threshold of your study, and prepares you in some measure to deal with the more complex problems of the living laboratory.

CHEMISTRY includes the art of separating and combining the elements of matter, and the study of the changes produced by these operations. We can hardly say too much of what it has contributed to our knowledge of the universe and our power of dealing with its materials. It has given us a *catalogue raisonné* of the substances found upon our planet, and shown how everything living and dead is put together from them. It is accomplishing wonders before us every day, such as Arabian story-tellers used to string together in their fables. It spreads the sensitive film on the artificial retina which looks upon us through the optician's lens for a few seconds, and fixes an image that will outlive its original. It questions the light of the sun, and detects the vaporized metals floating around the great luminary, — iron, sodium, lithium, and the rest, — as if the chemist of our remote planet could fill his bell-glasses from its fiery atmosphere.* It lends the power which flashes our messages in thrills that leave the lazy chariot of day behind them. It seals up a few dark grains in iron vases, and lo! at the touch of a single spark, rises in smoke and flame a mighty Afrit with a voice like thunder and an arm that shatters like an earthquake. The dreams of Oriental fancy have become the sober facts of our every-day life, and the chemist is the magician to whom we owe them.

To return to the colder scientific aspect of chemis-

* Scientific Annual for 1861. — FAIRBAIRN's Address before the British Association, 1861.

try. It has shown us how bodies stand affected to each other through an almost boundless range of combinations. It has given us a most ingenious theory to account for certain fixed relations in these combinations. It has successfully eliminated a great number of proximate compounds, more or less stable, from organic structures. It has *invented* others which form the basis of long series of well-known composite substances. In fact, we are perhaps becoming overburdened with our list of proximate principles, demonstrated and hypothetical.

How much nearer have we come to the secret of *force* than Lully and Geber and the whole crew of juggling alchemists? We have learned a great deal about the *how*, what have we learned about the *why?*

Why does iron rust, while gold remains untarnished, and gold amalgamate, while iron refuses the alliance of mercury?

The alchemists called gold Sol, the sun, and iron Mars, and pleased themselves with fancied relations between these substances and the heavenly bodies, by which they pretended to explain the facts they observed. Some of their superstitions have lingered in practical medicine to the present day, but chemistry has grown wise enough to confess the fact of absolute ignorance.

What is it that makes common salt crystallize in the form of cubes, and saltpetre in the shape of six-sided prisms? We see no reason why it should not have been just the other way, salt in prisms and saltpetre in cubes, or why either should take an exact geometrical outline, any more than coagulating albumen.

But although we had given up attempting to explain the essential nature of affinities and of crystalline types, we might have supposed that we had at least fixed the identity of the substances with which we deal, and determined the laws of their combination. All at once we find that a simple substance changes face, puts off its characteristic qualities and resumes them at will; — not merely when we liquefy or vaporize a solid, or reverse the process; but that a solid is literally transformed into another solid under our own eyes. We thought we knew phosphorus. We warm a portion of it sealed in an empty tube, for about a week. It has become a brown infusible substance, which does not shine in the dark nor oxidate in the air. We heat it to 500° F., and it becomes common phosphorus again. We transmute sulphur in the same singular way. Nature, you know, gives us carbon in the shape of coal and in that of the diamond. It is easy to call these changes by the name *allotropism*, but not the less do they confound our hasty generalizations.

These facts of allotropism have some corollaries connected with them rather startling to us of the nineteenth century. There may be other transmutations possible besides those of phosphorus and sulphur. When Dr. Prout, in 1840, talked about azote and carbon being "formed" in the living system, it was looked upon as one of those freaks of fancy to which philosophers, like other men, are subject. But when Professor Faraday, in 1851, says, at a meeting of the British Association, that " his hopes are in the direction of proving that bodies called simple were really compounds, and may be formed artificially as soon as we are masters of the laws influencing their combinations," — when he comes forward and says that he has tried experiments at transmutation, and means, if his life is spared, to try them again, — how can we be surprised at the popular story of 1861, that Louis Napoleon has established a gold-factory and is glutting the mints of Europe with bullion of his own making?

And so with reference to the law of combinations. The old maxim was, *Corpora non agunt nisi soluta.* If two substances, *a* and *b*, are enclosed in a glass vessel, *c*, we do not expect the glass to change them, unless *a* or *b* or the compound *a b* has the power of dissolving the glass. But if for *a* I take oxygen, for *b* hydrogen, and for *c* a piece of spongy platinum, I find the first two combine with the common signs of combustion and form water, the third in the mean time undergoing no perceptible change. It has played the part of the unwedded priest, who marries a pair without taking a fee or having any further relation with the parties. We call this *catalysis, catalytic action,* the *action of presence,* or by what learned name we choose. Give what name to it we will, it is a manifestation of power which crosses our established laws of combination at a very open angle of intersection. I think we may find an analogy for it in electrical induction, the disturbance of the equilibrium of the electricity of a body by the approach of a charged body to it, without interchange of electrical conditions between the two bodies. But an analogy is not an explanation, and why a few drops of yeast should change a saccharine mixture to carbonic acid and alcohol, — a little leaven leavening the whole lump, — not by combining with it, but by setting a movement at work, we not only cannot explain, but the fact is such an exception to the recognized laws of combination, that Liebig is unwilling to admit the new force at all to which Berzelius had given the name so generally accepted.

The phenomena of *isomerism*, or identity of composition and proportions of constituents with difference of qualities, and of isomorphism, or identity of form in crystals which have one element substituted for another, were equally surprises to science; and although the mechanism by which they are brought about can be to a certain extent explained by a reference to the hypothetical atoms of which the elements are constituted, yet this is only turning the difficulty into a fraction with an infinitesimal denominator and an infinite numerator.

So far we have studied the working of force and its seeming anomalies in purely chemical phenomena. But we soon find that chemical force is developed by various other physical agencies, — by heat, by light, by

electricity, by magnetism, by mechanical agencies ;
and, *vice versa*, that chemical action develops heat,
light, electricity, magnetism, mechanical force, as we
see in our matches, galvanic batteries, and explosive
compounds. Proceeding with our experiments, we
find that every kind of force is capable of producing
all other kinds, or, in Mr. Faraday's language, that
" the various forms under which the forces of matter
are made manifest have a common origin, or, in other
words, are so directly related and mutually dependent
that they are convertible one into another."

Out of this doctrine naturally springs that of the
conservation of force, so ably illustrated by Mr. Grove,
Dr. Carpenter, and Mr. Faraday. This idea is no
novelty, though it seems so at first sight. It was
maintained and disputed among the giants of philos-
ophy. Des Cartes and Leibnitz denied that any new
motion originated in nature, or that any ever ceased
to exist ; all motion being in a circle, passing from
one body to another, one losing what the other gained.
Newton, on the other hand, believed that new motions
were generated and existing ones destroyed. On the
first supposition, there is a fixed amount of force always circulating in the universe. On the second, the
total amount may be increasing or diminishing. You
will find in the Annual of Scientific Discovery for
1858 a very interesting lecture by Professor Helmholtz of Bonn, in which it is maintained that a certain portion of force is lost in every natural process,

being converted into unchangeable heat, so that the universe will come to a stand-still at last, all force passing into heat, and all heat into a state of equilibrium.

The doctrines of the convertibility or specific equivalence of the various forms of force, and of its conservation, which is its logical consequence, are very generally accepted, as I believe, at the present time, among physicists. We are naturally led to the question, What is the nature of force? The three illustrious philosophers just referred to agree in attributing the general movements of the universe to the immediate Divine action.* The doctrine of "pre-established harmony" was an especial contrivance of Leibnitz to remove the Creator from unworthy association with the less divine acts of living beings. Obsolete as this expression sounds to our ears, the phrase *laws of the universe*, which we use so contantly with a wider application, appears to me essentially identical with it.

Force does not admit of explanation, nor of proper

---

* " Et generalem quod attinet, manifestum mihi videtur illam [causam] non aliam esse, quam Deum ipsum, qui materiam simul cum motu et quiete in principio creavit, jamque per solum suum concursum ordinarium, tantundem motus et quietis in ea tota quantum tunc posuit conservat: . . . . eodem plane modo, eademque ratione qua prius creavit, eum etiam tantundem motus in ipsa semper conservare." — DES CARTES, Princ. Phil., P. II. § XXXVI.

" Concursus Dei, actioni creaturæ necessarius." — LEIBNITZ, Op., Tom. VI. p. 174.

" In ipso continentur et moventur universa, sed absque mutua *passione.* Deus nihil patitur ex corporum motibus: illa nullam sentiunt resistentiam ex omnipræsentia Dei." — NEWTON, Principia, Lib. III. Schol. Gen.

definition, any more than the hypothetical substratum
of matter.   If we assume the Infinite as omnipresent,
omniscient, omnipotent, we cannot suppose Him ex-
cluded from any part of His creation, except from
rebellious souls which voluntarily exclude Him by
the exercise of their fatal prerogative of free-will.*
Force, then, is the act of immanent Divinity. I find
no meaning in mechanical explanations. Newton's hy-
pothesis of an ether filling the heavenly spaces does
not, I confess, help my conceptions. I will, and the
muscles of my vocal organs shape my speech. God
wills, and the universe articulates His power, wisdom,
and goodness. That is all I know. There is no bridge
my mind can throw from the "immaterial" cause to
the "material" effect.

The problem of force meets us everywhere, and I
prefer to encounter it in the world of physical phenomena before reaching that of living actions. It is only
the name for the incomprehensible cause of certain
changes known to our consciousness, and assumed to be
outside of it. For me it is the Deity Himself in action.
I can therefore see a large significance in the some-

---

* "Cum unaquæque spatii particula sit *semper*, et unumquodque duratio-
nis indivisibile momentum *ubique*; certe rerum omnium Fabricator ac
Dominus non erit *nunquam nusquam*. Omnipræsens est non per *virtutem*
solam, sed etiam per *substantiam*; nam virtus sine substantia subsistere
non potest." — NEWTON, *loc. cit.*

    " The Lord of all, himself throûgh all diffused,
      Sustains and is the life of all that lives."
                                    The Task, B. VI. l. 221, 222.

what bold language of Burdach: " There is for me but one miracle, that of infinite existence, and but one mystery, the manner in which the finite proceeds from the infinite. So soon as we recognize this incomprehensible act as the general and primordial miracle, of which our reason perceives the necessity, but the manner of which our intelligence cannot grasp, so soon as we contemplate the nature known to us by experience, in this light, there is for us no other impenetrable miracle or mystery." *

Let us turn to a branch of knowledge which deals with certainties up to the limit of the senses, and is involved in no speculations beyond them. In certain points of view, HUMAN ANATOMY may be considered an almost exhausted science. From time to time some small organ which had escaped earlier observers has been pointed out, — such parts as the *tensor tarsi*, the otic ganglion, or the Pacinian bodies; but some of our best anatomical works are those which have been classic for many generations. The plates of the bones in Vesalius, three centuries old, are still masterpieces of accuracy, as of art. The magnificent work of Albinus on the muscles, published in 1747, is still supreme in its department, as the constant references of the most thorough recent treatise on the subject, that of Theile, sufficiently show. More has been done in unravelling the mysteries of the fasciæ, but there has

* Physiologie, Trad. de Jourdan, II. 326.

B

been a tendency to overdo this kind of material analysis. Alexander Thomson split them up into cobwebs,
as you may see in the plates to Velpeau's Surgical
Anatomy. I well remember how he used to shake
his head over the coarse work of Scarpa and Astley
Cooper,—as if Denner, who painted the separate hairs
of the beard and pores of the skin in his portraits, had
spoken lightly of the pictures of Rubens and Vandyk.

Not only has little been added to the catalogue of
parts, but some things long known had become half-
forgotten. Louis and others confounded the solitary
glands of the lower part of the small intestine with
those which "the great Brunner," as Haller calls him,
described in 1687 as being found in the duodenum.
The display of the fibrous structure of the brain seemed
a novelty as shown by Spurzheim. One is startled to
find the method anticipated by Raymond Vieussens
nearly two centuries ago. I can hardly think Gordon
had ever looked at his figures, though he names their
author, when he wrote the captious and sneering article which attracted so much attention in the pages of
the Edinburgh Review.*

This is the place, if anywhere, to mention any observations I could pretend to have made in the course
of my teaching the structure of the human body. I
can make no better show than most of my predecessors
in this well-reaped field. The nucleated cells found
connected with the cancellated structure of the bones,

* June, 1815.

which I first pointed out and had figured in 1847, and have shown yearly from that time to the present, and the *fossa masseterica*, a shallow concavity on the ramus of the lower jaw, for the lodgement of the masseter muscle, which acquires significance when examined by the side of the deep cavity on the corresponding part in some carnivora to which it answers, may perhaps be claimed as deserving attention. I have also pleased myself by making a special group of the six radiating muscles * which diverge from the spine of the axis, or second cervical vertebra, and by giving to it the name *stella musculosa nuchœ*. But this scanty catalogue is only an evidence that one may teach long and see little that has not been noted by those who have gone before him. Of course I do not think it necessary to include rare, but already described anomalies, such as the episternal bones, the rectus sternalis, and other interesting exceptional formations I have encountered, which have shown a curious tendency to present themselves several times in the same season, perhaps because the first specimen found calls our attention to any we may subsequently meet with.

The anatomy of the scalpel and the amphitheatre was, then, becoming an exhausted branch of investigation. But during the present century the study of the human body has changed its old aspect, and become fertile in new observations. This rejuvenescence was

---

* *Rectus capitis posticus major, obliquus capitis inferior*, and *semispinalis colli*, on each side.

effected by means of two principal agencies, — new methods and a new instrument.

Descriptive anatomy, as known from an early date, is to the body what geography is to the planet. Now geography was pretty well known so long ago as when Arrowsmith, who was born in 1750, published his admirable maps. But in that same year was born Werner, who taught a new way of studying the earth, since become familiar to us all under the name of *Geology.*

What geology has done for our knowledge of the earth, has been done for our knowledge of the body by that method of study to which is given the name of *General Anatomy.* It studies, not the organs as such, but the elements out of which the organs are constructed. It is the geology of the body, as that is the general anatomy of the earth. The extraordinary genius of Bichat, to whom more than any other we owe this new method of study, does not require Mr. Buckle's testimony to impress the practitioner with the importance of its achievements. I have heard a very wise physician question whether any important result had accrued to practical medicine from Harvey's discovery of the circulation. But Anatomy, Physiology, and Pathology have received a new light from this novel method of contemplating the living structures, which has had a vast influence in enabling the practitioner at least to distinguish and predict the course of disease. We know as well what differences to expect in the habits of a mucous and of a serous membrane, as what mineral substances to look for in the chalk or the coal measures. You have only to read Cullen's description of inflammation of the lungs or of the bowels, and compare it with such as you may find in Laennec or Watson, to see the immense gain which diagnosis and prognosis have derived from general anatomy.

The second new method of studying the human structure, beginning with the labors of Scarpa, Burns, and Colles, grew up principally during the first third of this century. It does not deal with organs, as did the earlier anatomists, nor with tissues, after the manner of Bichat. It maps the whole surface of the body into an arbitrary number of regions, and studies each region successively from the surface to the bone, or beneath it. This hardly deserves the name of a science, although Velpeau has dignified it with that title, but it furnishes an admirable practical way for the surgeon who has to operate on a particular region of the body to study that region. If we are buying a farm, we are not content with the State map or a geological chart including the estate in question. We demand an exact survey of that particular property, so that we may know what we are dealing with. This is just what regional, or, as it is sometimes called, surgical anatomy, does for the surgeon with reference to the part on which his skill is to be exercised. It enables him to see with the mind's eye through the opaque tissues down to the bone on which they lie, as if the skin were transparent as the cornea, and the organs it covers translucent as the gelatinous pulp of a medusa.

It is curious that the Japanese should have anticipated Europe in a kind of rude regional anatomy. I have seen a manikin of Japanese make traced all over with lines, and points marking their intersection. By this their doctors are guided in the performance of acupuncture, marking the safe places to thrust in needles, as we buoy out our ship-channels, and doubtless indicating to learned eyes the spots where incautious meddling had led to those little accidents of shipwreck to which patients are unfortunately liable.

A change of method, then, has given us General and Regional Anatomy. These, too, have been worked so thoroughly, that, if not exhausted, they have at least become to a great extent fixed and positive branches of knowledge. But the first of them, General Anatomy, would never have reached this positive condition but for the introduction of that instrument which I have mentioned as the second great aid to modern progress.

This instrument is the achromatic microscope. For the history of the successive steps by which it became the effective scientific implement we now possess, I must refer you to the work of Mr. Quekett, to an excellent article in the Penny Cyclopædia, or to that of Sir David Brewster in the Encyclopædia Britannica. It is a most interesting piece of scientific history, which shows how the problem that Biot in 1821 pronounced insolvable was in the course of a few years practically solved, with a success equal to that which Dollond had long before obtained with the telescope. It is enough for our purpose that we are now in possession of an instrument freed from all confusions and illusions, which magnifies a thousand diameters, — a million times in surface, — without serious distortion or discoloration of its object.

A quarter of a century ago, or a little more, an instructor would not have hesitated to put John Bell's Anatomy and Bostock's Physiology into a student's hands, as good authority on their respective subjects. Let us not be unjust to either of these authors.   John Bell is the liveliest medical writer that I can remember who has written since the days of delightful old Ambroise Paré.   His picturesque descriptions and bold figures are as good now as they ever were, and his book can never become obsolete.   But listen to what John Bell says of the microscope: —

"Philosophers of the last age had been at infinite pains to find the ultimate fibre of muscles, thinking to discover its properties in its form ; but they saw just in proportion to the glasses which they used, or to their practice and skill in that art, which is now almost forsaken." *

Dr. Bostock's work, neglected as it is, is one which I value very highly as a really learned compilation, full

* Anat. and Phys. of the Human Body, I. 273.

of original references.   But Dr. Bostock says: " Much
as the naturalist has been indebted to the microscope,
by bringing into view many beings of which he could
not otherwise have ascertained the existence, the physi-
ologist has not yet derived any great benefit from the
instrument." *

These are only specimens of the manner in which
the microscope and its results were generally regarded
by the generation just preceding our own.

I have referred you to the proper authorities for the
account of those improvements which about the year
1830 rendered the compound microscope an efficient
and trustworthy instrument.   It was now for the first
time that a true general anatomy became possible.   As
early as 1816 Treviranus had attempted to resolve the
tissues, of which Bichat had admitted no less than
twenty-one, into their simple microscopic elements.
How could such an attempt succeed, Henle well
asks,† at a time when the most extensively diffused
of all the tissues, the areolar, was not at all understood?   All that method could do had been accomplished by Bichat and his followers.   It was for the
optician to take the next step.   The future of anatomy
and physiology, as an enthusiastic micrologist of the
time said, was in the hands of Messrs. Schieck and
Pistor, famous opticians of Berlin.

In those earlier days of which I am speaking, all

* Physiology, p. 281.
† Anatomie Générale, (Trad. de Jourdan,) I. 125.

the points of minute anatomy were involved in obscurity. Some found globules everywhere, some fibres. Students disputed whether the conjunctiva extended over the cornea or not, and worried themselves over Gaultier de Claubry's stratified layers of the skin, or Breschet's blennogenous and chromatogenous organs. The dartos was a puzzle, the central spinal canal a myth, the decidua clothed in fable as much as the golden fleece. The structure of bone, now so beautifully made out, — even that of the teeth, in which old Leeuwenhoek, peeping with his octogenarian eyes through the minute lenses wrought with his own hands, had long ago seen the "pipes," as he called them, — was hardly known at all. The minute structure of the viscera lay in the mists of an uncertain microscopic vision. The intimate recesses of the animal system were to the students of anatomy what the interior of Africa long was to geographers, and the stories of microscopic explorers were as much sneered at as those of Bruce or Du Chaillu, and with better reason.

Now what have we come to in our own day? In the first place, the minute structure of all the organs has been made out in the most satisfactory way. The special arrangements of the vessels and the ducts of all the glands, of the air-tubes and vesicles of the lungs, of the parts which make up the skin and other membranes, all the details of those complex parenchymatous organs which had confounded investigation so long, have been lifted out of the invisible into the sight of all

2

observers. It is fair to mention here, that we owe a
great deal to the art of minute injection, by which we
are enabled to trace the smallest vessels in the midst
of the tissues where they are distributed. This is an
old artifice of anatomists. The famous Ruysch, who
died a hundred and thirty years ago, showed that each
of the viscera has its terminal vessels arranged in its
own peculiar way; * the same fact which you may see
illustrated in Gerber's figures after the minute injec-
tions of Berres.† I hope to show you many specimens
of this kind in the microscope, the work of English and
American hands. Professor Agassiz allows me also to
make use of a very rich collection of injected preparations sent him by Professor Hyrtl, formerly of Prague,
now of Vienna, for the proper exhibition of which I
had a number of microscopes made expressly, by Mr.
Grunow, during the past season. All this illustrates
what has been done for the elucidation of the intimate
details of formation of the organs.

But the great triumph of the microscope as applied
to anatomy has been in the resolution of the organs
and the tissues into their simple constituent anatomical
elements. It has taken up general anatomy where
Bichat left it. He had succeeded in reducing the
structural language of nature to syllables, if you will
permit me to use so bold an image. The microscopic
observers who have come after him have analyzed these

---

* Haller, Bibl. Anat., I. 533.
† General and Minute Anatomy, (London, 1842,) Plate XXIII.

into *letters*, as we may call them,—the simple elements by the combination of which Nature spells out successively tissues, which are her syllables, organs which are her words, systems which are her chapters, and so goes on from the simple to the complex, until she binds up in one living whole that wondrous volume of power and wisdom which we call the human body.

The alphabet of the organization is so short and simple, that I will risk fatiguing your attention by repeating it, according to the plan I have long adopted.

A. Cells, either floating, as in the blood, or fixed, like those in the cancellated structure of bone, already referred to. Very commonly they have undergone a change of figure, most frequently a flattening which reduces them to scales, as in the epidermis and the epithelium.

B. Simple, translucent, homogeneous solid, such as is found at the back of the cornea, or forming the intercellular substance of cartilage.

C. The white fibrous element, consisting of very delicate, tenacious threads. This is the long-staple textile substance of the body. It is to the organism what cotton is pretended to be to our Southern States. It pervades the whole animal fabric as areolar tissue, which is the universal packing and wrapping material. It forms the ligaments which bind the whole framework together. It furnishes the sinews, which are the channels of power. It enfolds every muscle. It wraps the brain in its hard, insensible folds, and the heart itself beats in a purse that is made of it.

D. The yellow elastic, fibrous element, the caout-
chouc of the animal mechanism, which pulls things
back into place, as the india-rubber band shuts the
door we have opened.

E. The striped muscular fibre, — the red flesh,
which shortens itself in obedience to the will, and thus
produces all voluntary active motion.

F. The unstriped muscular fibre, more properly
the fusiform-cell fibre, which carries on the involuntary internal movements.

G. The nerve-cylinder, a glassy tube, with a pith
of some firmness, which conveys sensation to the brain
and the principle which induces motion from it.

H. The nerve-corpuscle, the centre of nervous
power.

I. The mucous tissue, as Virchow calls it, common
in embryonic structures, seen in the vitreous humor of
the adult.

To these add X, granules, of indeterminate shape
and size, Y, for inorganic matters, such as the salts of
bone and teeth, and Z, to stand as a symbol of the
fluids, and you have the letters of what I have ventured to call the alphabet of the body.

But just as in language certain diphthongs and syllables are frequently recurring, so we have in the body
certain secondary and tertiary combinations, which we
meet more frequently than the solitary elements of
which they are composed.

Thus A B, or a collection of cells united by simple

structureless solid, is seen to be extensively employed in the body under the name of *cartilage*. Out of this the surfaces of the articulations and the springs of the breathing apparatus are formed. But when Nature came to the buffers of the spinal column (interverte-bral disks) and the washers of the joints (semilunar fibro-cartilages of the knee, etc.), she required more tenacity than common cartilage possessed. What did she do? What does man do in a similar case of need? I need hardly tell you. The mason lays his bricks in simple mortar. But the plasterer works some *hair* into the mortar which he is going to lay in large sheets on the walls. The children of Israel complained that they had no *straw* to make their bricks with, though portions of it may still be seen in the crumbling pyra-mid of Darshour, which they are said to have built. I visited the old house on Witch Hill in Salem a year or two ago, and there I found the walls coated with clay in which straw was abundantly mingled; — the old Judaizing witch-hangers copied the Israelites in a good many things. The Chinese and the Corsicans blend the fibres of amianthus in their pottery to give it tenacity. Now to return to Nature. To make her buffers and washers hold together in the shocks to which they would be subjected, she took common car-tilage and mingled the white fibrous tissue with it, to serve the same purpose as the hair in the mortar, the straw in the bricks and in the plaster of the old wall, and the amianthus in the earthen vessels. Thus we

have the combination A B C, or *fibro-cartilage*. Again,
the bones were once only gristle or cartilage, A B.
To give them solidity they were infiltrated with stone,
in the form of salts of lime, an inorganic element, so
that bone would be spelt out by the letters A, B, and Y.

If from these organic syllables we proceed to form
organic words, we shall find that Nature employs
three principal forms; namely, Vessels, Membranes,
and Parenchyma, or visceral tissue. The most complex of them can be resolved into a combination of
these few simple anatomical constituents.

Passing for a moment into the domain of PATHOLOGI-
CAL ANATOMY, we find the same elements in morbid
growths that we have met with in normal structures.
The pus-corpuscle and the white blood-corpuscle can
only be distinguished by tracing them to their origin.
A frequent form of so-called malignant disease proves
to be only a collection of altered epithelium-cells.
Even cancer itself has no specific anatomical element, and the diagnosis of a cancerous tumor by the
microscope, though tolerably sure under the eye of
an expert, is based upon accidental, and not essential
points, — the crowding together of the elements, the
size of the cell-nuclei, and similar variable characters.

Let us turn to PHYSIOLOGY. The microscope, which
has made a new science of the intimate structure of
the organs, has at the same time cleared up many

uncertainties concerning the mechanism of the special functions. Up to the time of the living generation of observers, Nature had kept over all her inner workshops the forbidding inscription, *No Admittance!* If any prying observer ventured to spy through his magnifying tubes into the mysteries of her glands and canals and fluids, she covered up her work in blinding mists and bewildering halos, as the deities of old concealed their favored heroes in the moment of danger. Science has at length sifted the turbid light of her lenses, and blanched their delusive rainbows.

Anatomy studies the organism in space. Physiology studies it also in time. After the study of form and composition follows close that of action, and this leads us along back to the first moment of the germ, and forward to the resolution of the living frame into its lifeless elements. In this way Anatomy, or rather that branch of it which we call Histology, has become inseparably blended with the study of function. The connection between the science of life and that of intimate structure on the one hand, and composition on the other, is illustrated in the titles of two recent works of remarkable excellence, — the Physiological Anatomy of Todd and Bowman, and the Physiological Chemistry of Lehmann.

Let me briefly recapitulate a few of our acquisitions in Physiology, due in large measure to our new instruments and methods of research, and at the same time indicate the limits which form the permanent or the

temporary boundaries of our knowledge.   I will begin
with the largest fact and with the most absolute and
universally encountered limitation.

The "largest truth in Physiology" Mr. Paget con-
siders to be "the development of ova through multi-
plication and division of their cells."   I would state
it more broadly as the agency of the cell in all living
processes.   It seems at present necessary to abandon
the original idea of Schwann, that we can observe the
building up of a cell from the simple granules of a
blastema, or formative fluid.   The evidence points
rather towards the axiom, *Omnis cellula e cellula ;*
that is, the germ of a new cell is always derived from
a pre-existing cell.   The doctrine of Schwann, as I remarked long ago (1844), runs parallel with the nebu-
lar theory in astronomy, and they may yet stand or fall
together.

As we have seen Nature anticipating the plasterer
in fibro-cartilage, so we see her beforehand with the
glass-blower in her dealings with the cell.   The artisan
blows his vitreous bubbles, large or small, to be used
afterwards as may be wanted.   So Nature shapes her
hyaline vesicles and modifies them to serve the needs
of the part where they are found.   The artisan whirls
his rod, and his glass bubble becomes a flattened disk,
with its bull's-eye for a nucleus.   These lips of ours
are all glazed with microscopic tiles formed of flattened
cells, each one of them with its nucleus still as plain
and relatively as prominent, to the eye of the micro-

scopist, as the bull's-eye in the old-fashioned window-pane. Everywhere we find cells, modified or un-changed. They roll in inconceivable multitudes (five millions and more to the cubic millimetre, according to Vierordt*) as blood-disks through our vessels. A close-fitting mail of flattened cells coats our surface with a panoply of imbricated scales, (more than twelve thousand millions, as Harting has computed,†) as true a defence against our enemies as the buckler of the armadillo or the carapace of the tortoise against theirs. The same little protecting organs pave all the great highways of the interior system. Cells, again, preside over the chemical processes which elaborate the living fluids; they change their form to become the agents of voluntary and involuntary motion; the soul itself sits on a throne of nucleated cells, and flashes its mandates through skeins of glassy filaments which once were simple chains of vesicles. And, as if to reduce the problem of living force to its simplest expression, we see the yolk of a transparent egg dividing itself in whole or in part, and again dividing and subdividing, until it becomes a mass of cells, out of which the harmonious diversity of the organs arranges itself, worm or man, as God has willed from the beginning.

This differentiation having been effected, each several part assumes its special office, having a life of its own, adjusted to that of other parts and the whole.

* Kölliker, Manual, etc., (London, 1860,) p. 518.
† Valentin's Physiology, (Brinton's Transl.,) p. 13.

2*

c

"Just as a tree constitutes a mass arranged in a definite manner, in which, in every single part, in the leaves as in the root, in the trunk as in the blossom, cells are discovered to be the ultimate elements, so is it also with the forms of animal life. *Every animal presents itself as a sum of vital unities*, every one of which manifests all the characteristics of life." *

The *mechanism* is as clear, as unquestionable, as absolutely settled and universally accepted, as the order of movement of the heavenly bodies, which we compute backward to the days of the observatories on the plains of Shinar, and on the faith of which we regulate the movements of war and trade by the predictious of our ephemeris.

The *mechanism*, and that is all. We see the workman and the tools, but the skill that guides the work and the power that performs it are as invisible as ever. I fear that not every listener took the significance of those pregnant words in the passage I quoted from John Bell, — "*thinking to discover its properties in its form.*" We have discovered the working bee in this great hive of organization. We have detected the cell in the very act of forming itself from a nucleus, of transforming itself into various tissues, of selecting the elements of various secretions. But why one cell becomes nerve and another muscle, why one selects bile and another fat, we can no more pretend to tell, than why one grape sucks out of the soil the generous juice

* Virchow, Cellular Pathology, Lect. I.

which princes hoard in their cellars, and another the wine which it takes three men to drink, — one to pour it down, another to swallow it, and a third to hold him while it is going down. Certain analogies between this selecting power and the phenomena of endosmosis in the elective affinities of chemistry we can find, but the problem of force remains here, as everywhere, unsolved and insolvable.

Do we gain anything by attempting to get rid of the idea of a special vital force because we find certain mutually convertible relations between forces in the body and out of it? I think not, any more than we should gain by getting rid of the idea and expression Magnetism because of its correlation with electricity. We may concede the unity of all forms of force, but we cannot overlook the fixed differences of its manifestations according to the conditions under which it acts. It is a mistake, however, to think the mystery is greater in an organized body than in any other. We see a stone fall or a crystal form, and there is nothing stranger left to wonder at, for we have seen the Infinite in action.

Just so far as we can recognize the ordinary modes of operation of the common forces of nature, — gravity, cohesion, elasticity, transudation, chemical action, and the rest, — we see the so-called vital acts in the light of a larger range of known facts and familiar analogies. Matteucci's well-remembered lectures contain many and striking examples of the working of physical forces

in physiological processes. Wherever rigid experiment
carries us, we are safe in following this lead ; but the
moment we begin to theorize beyond our strict observation, we are in danger of falling into those mechanical
follies which true science has long outgrown.

Recognizing the fact, then, that we have learned
nothing but the machinery of life, and are no nearer
to its essence, what is it that we have gained by this
great discovery of the cell formation and function ?

It would have been reward enough to learn the
method Nature pursues for its own sake. If the sov-
ereign Artificer lets us into his own laboratories and
workshops, we need not ask more than the privilege
of looking on at his work. We do not know where
we now stand in the hierarchy of created intelligences.
We were *made* a little lower than the angels. I speak
it not irreverently ; as the lower animals surpass man
in some of their attributes, so it may be that not every
angel's eye can see as broadly and as deeply into the
material works of God as man himself, looking at the
firmament through an equatorial of fifteen inches' aper-
ture, and searching into the tissues with a twelfth
of an inch objective.

But there are other positive gains of a more practical character. Thus we are no longer permitted to
place the seat of the living actions in the extreme vessels, which are only the carriers from which each part
takes what it wants by the divine right of the omnipotent nucleated cell. The organism has become, in the

words already borrowed from Virchow, " a sum of vital unities." The *strictum* and *laxum*, the increased and diminished action of the vessels, out of which medical theories and methods of treatment have grown up, have yielded to the doctrine of local cell-communities, belonging to this or that vascular district, from which they help themselves, as contractors are wont to do from the national treasury.

I cannot promise to do more than to select a few of the points of contact between our ignorance and our knowledge which present particular interest in the existing state of physiological knowledge. Some of them involve the microscopic discoveries of which I have been speaking, some belong to the domain of chemistry, and some have relations with other departments of physical science.

If we should begin with the digestive function, we should find that the long-agitated question of the nature of the acid of the gastric juice is becoming settled in favor of the lactic. But the whole solvent agency of the digestive fluid enters into the category of that exceptional mode of action already familiar to us in chemistry as catalysis. It is therefore doubly difficult of explanation ; first, as being, like all reactions, a fact not to be accounted for except by the imaginative appeal to " affinity," and secondly, as being one of those peculiar reactions provoked by an element which stands outside and looks on without compromising itself.

The doctrine of Mulder, so widely diffused in popular and scientific belief, of the existence of a common base of all albuminous substances, the so-called *protein*, has not stood the test of rigorous analysis. The division of food into azotized and non-azotized is no doubt important, but the attempt to show that the first only is *plastic* or nutritive, while the second is simply *calorifacient*, or heat-producing, fails entirely in the face of the facts revealed by the study of man in different climates, and of numerous experiments in the feeding of animals. I must return to this subject in connection with the respiratory function.

The sugar-making faculty of the liver is another "catalytic" mystery, as great as the rest of them, and no greater. Liver-tissue brings sugar out of the blood, or out of its own substance; — why ?

> *Quia est in eo*
> *Virtus saccharitiva.*

Just what becomes of the sugar beyond the fact of its disappearance before it can get into the general circulation and sweeten our tempers, it is hard to say.

The pancreatic fluid makes an emulsion of the fat contained in our food, but just how the fatty particles get into the villi we must leave Brücke and Kölliker to settle if they can.

No one has shown satisfactorily the process by which the blood-corpuscles are formed out of the lymph-corpuscles, nor what becomes of them. These two questions are like those famous household puzzles, — Where do the flies come from? and, Where do the pins go to?

There is a series of organs in the body which has long puzzled physiologists, — organs of glandular aspect, but having no ducts, — the spleen, the thyroid and thymus bodies, and the suprarenal capsules. We call them *vascular glands*, and we believe that they elaborate colored and uncolored blood-cells ; but just what changes they effect, and just how they effect them, it has proved a very difficult matter to determine. So of the noted glandules which form Peyer's patches, their precise office, though seemingly like those of the lymphatic glands, cannot be positively assigned, so far as I know, at the present time. It is of obvious interest to learn it with reference to the pathology of typhoid fever. It will be remarked that the coincidence of their changes in this disease with enlargement of the spleen suggests the idea of a similarity of function in these two organs.

The theories of the production of animal heat, from the times of Black, Lavoisier, and Crawford to those of Liebig, are familiar to all who have paid any attention to physiological studies. The simplicity of Liebig's views, and the popular form in which they have been presented, have given them wide currency, and incorporated them in the common belief and language of our text-books. Direct oxidation or combustion of the carbon and hydrogen contained in the food, or in the tissues themselves ; the division of alimentary substances into *respiratory*, or non-azotized, and azotized, — these doctrines are familiar even to the classes in our high-schools. But this simple statement is boldly ques-

tioned. Nothing proves that oxygen combines (in the
system) with hydrogen and carbon in particular, rather
than with sulphur and azote. Such is the well-grounded
statement of Robin and Verdeil. " It is very probable
that animal heat is entirely produced *by the chemical
actions which take place in the organism*, but the phenomenon is too complex to admit of our calculating it
according to the quality of oxygen consumed." These
last are the words of Regnault, as cited by Mr. Lewes,
whose intelligent discussion of this and many of the
most interesting physiological problems I strongly recommend to your attention.

This single illustration covers a wider ground than
the special function to which it belongs. We are learning that the chemistry of the body must be studied, not
simply by its *ingesta* and *egesta*, but that there is a
long intermediate series of changes which must be in-
vestigated in their own light, under their own special
conditions. The expression " sum of vital unities "
applies to the chemical actions, as well as to other
actions localized in special parts ; and when the distinguished chemists whom I have just cited entitle
their work a treatise on the *immediate principles* of the
body, they only indicate the nature of that profound
and subtile analysis which must take the place of all
hasty generalizations founded on a comparison of the
food with residual products.

I will only call your attention to the fact, that the
exceptional phenomenon of the laboratory is the pre-

vailing law of the organism. Nutrition itself is but one
great catalytic process. As the blood travels its rounds,
each part selects its appropriate element and transforms
it to its own likeness. Whether the appropriating agent
be cell or nucleus, or a structureless solid like the intercellular substance of cartilage, the fact of its presence determines the separation of its proper constituents from the circulating fluid, so that even when we
are wounded bone is replaced by bone, skin by skin,
and nerve by nerve.

It is hardly without a smile that we resuscitate the
old question of the *vis insita* of the muscular fibre, so
famous in the discussions of Haller and his contemporaries. Speaking generally, I think we may say that
Haller's doctrine is the one now commonly received ;
namely, that the muscles contract in virtue of their
own inherent endowments. It is true that Kölliker
says no perfectly decisive fact has been brought forward to prove that the *striated* muscles contract without having been acted on by nerves. Yet Mr. Bowman's observations on the contraction of isolated fibres
appear decisive enough (unless we consider them inval-
idated by Dr. Lionel Beale's recent researches, tending
to show that each elementary fibre is supplied with
nerves *) ; and as to the smooth muscular fibres, we
have Virchow's statement respecting the contractility

* Proc. Royal Society, No. XL. Vol. X., and British and Foreign Med.
Chir. Review for April, 1861.

of those of the umbilical cord, where there is not a trace of any nerves.*

In the investigation of the nervous system, anatomy and physiology have gone hand in hand. It is very singular that so important, and seemingly simple, a fact as the connection of the nerve-tubes, at their origin or in their course, with the nerve-cells, should have so long remained open to doubt, as you may see that it did by referring to the very complete work of Sharpey and Quain, (edition of 1849,) the histological portion of which is cordially approved by Kölliker himself.†

Several most interesting points of the minute anatomy of the nervous centres have been laboriously and skilfully worked out by a recent graduate of this Medical School, in a monograph worthy to stand in line with those of Lockhart Clarke, Stilling, and Schröder van der Kolk.‡ I have had the privilege of examining and of showing some of you a number of Dr. Dean's skilful preparations. I have no space to give even an abstract of his conclusions. I can

---

* See also the results of experiments with *woorara* and *sulphocyanide of potassium*. The first destroys the irritability of the nerves, the second that of the muscles. The student will find a notice of Bernard's experiments with these poisons in Dr. Dalton's standard work on Physiology, which if he does not own, he should at once procure.

† See also a learned note in Dr. Waldo I. Burnett's " Reviews and Abstracts," etc., American Journal of Science, September, 1853.

‡ Microscopic Anatomy of the Lumbar Enlargement of the Spinal Cord. By John Dean, M. D. Cambridge. 1861.

only refer to his proof of the fact, that a single cell may send its processes into several different bundles of nerve-roots (Fig. 7, *B*), and to his demonstration of the curved ascending and descending fibres from the posterior nerve-roots, to reach what he has called the *longitudinal columns of the cornua* (Fig. 8, *h*, *h*). I must also mention Dr. Dean's exquisite microscopic photographs from sections of the medulla oblongata, which appear to me to promise a new development, if not a new epoch, in anatomical art.

It having been settled that the nerve-tubes can very commonly be traced directly to the nerve-cells, the object of all the observers in this department of anatomy is to follow these tubes to their origin. We have an infinite snarl of telegraph-wires, and we may be reasonably sure that, if we can follow them up, we shall find each of them ends in a battery somewhere. One of the most interesting problems is to find the ganglionic origin of the great nerves of the medulla oblongata, and this is the end to which, by the aid of the most delicate sections, colored so as to bring out their details, mounted so as to be imperishable, magnified by the best instruments, and now self-recorded in the light of the truth-telling sunbeam, our fellow-student is making a steady progress in a labor which I think bids fair to rank with the most valuable contributions to histology that we have had from this side of the Atlantic.

It is interesting to see how old questions are incidentally settled in the course of these new investigations. Thus, Mr. Clarke's dissections, confirmed by preparations of Mr. Dean's which I have myself examined, place the fact of the decussation of the pyramids — denied by Haller, by Morgagni, and even by Stilling — beyond doubt. So the spinal canal, the existence of which, at least in the adult, has been so often disputed, appears as a coarse and unequivocal anatomical fact in many of the preparations referred to.

While these studies of the structure of the cord have been going on, the ingenious and indefatigable Brown-Séquard has been investigating the functions of its different parts with equal diligence. The microscopic anatomists had shown that the ganglionic corpuscles of the gray matter of the cord are connected with each other by their processes, as well as with the nerve-roots. M. Brown-Séquard has proved by numerous experiments that the gray substance transmits sensitive impressions and muscular stimulation. The oblique ascending and descending fibres from the posterior nerve-roots, joining the " longitudinal columns of the cornua," * account for the results of Brown-Séquard's sections of the posterior columns.† The physiological experimenter has also made it evident that the decussation of the conductors of sensitive impressions has its seat in the spinal cord, and not in the encephalon, as had been supposed. Not

* Dean's Memoir, Fig. 8.
† Lectures, (Philadelphia, 1860,) Lect. II. p. 26, and Plate I. fig. 7.

less remarkable than these results are the facts, which I with others of my audience have had the opportunity of observing, as shown by M. Brown-Séquard, of the artificial production of epilepsy in animals by injuring the spinal cord, and the induction of the paroxysm by pinching a certain portion of the skin. I would also call the student's attention to his account of the relations of the nervous centres to nutrition and secretion, the last of which relations has been made the subject of an extended essay by our fellow-countryman, Dr. H. F. Campbell, of Georgia.

The physiology of the spinal cord seems a simple matter as you study it in Longet. The experiments of Brown-Séquard have shown the problem to be a complex one, and raised almost as many doubts as they have solved questions; at any rate, I believe all lecturers on physiology agree that there is no part of their task they dread so much as the analysis of the evidence relating to the special offices of the different portions of the medulla spinalis. In the brain we are sure that we do not know how to localize functions; in the spinal cord, we think we do know something; but there are so many anomalies, and seeming contradictions, and sources of fallacy, that, beyond the facts of crossed paralysis of sensation, and the conducting' agency of the gray substance, I am afraid we retain no cardinal principles discovered since the development of the reflex function took its place by Sir Charles Bell's great discovery.

By the manner in which I spoke of the brain, you will see that I am obliged to leave phrenology *sub Jove*, — out in the cold, — as not one of the household of science. I am not one of its haters ; on the contrary, I am grateful for the incidental good it has done. I love to amuse myself in its plaster Golgothas, and listen to the glib professor, as he discovers by his manipulations

<blockquote>" All that disgraced my betters met in me."</blockquote>

I loved of old to see square-headed, heavy-jawed Spurzheim make a brain flower out into a corolla of marrowy filaments, as Vieussens had done before him, and to hear the dry-fibred but human-hearted George Combe teach good sense under the disguise of his equivocal system. But the pseudo-sciences, phrenology and the rest, seem to me only appeals to weak minds and the weak points of strong ones. There is a *pica* or false appetite in many intelligences ; they take to odd fancies in place of wholesome truth, as girls gnaw at chalk and charcoal. Phrenology juggles with nature. It is so adjusted as to soak up all evidence that helps it, and shed all that harms it. It crawls forward in all weathers, like Richard Edgeworth's hygrometer. It does not stand at the boundary of our ignorance, it seems to me, but is one of the will-o'-the-wisps of its undisputed central domain of bog and quicksand. Yet I should not have devoted so many words to it, did I not recognize the light it has thrown on human actions by its study of congenital organic tendencies. Its maps of the surface of the head are, I feel sure, founded on a delusion, but its studies of individual character are always interesting and instructive.

The "snapping-turtle" strikes after its natural fashion when it first comes out of the egg. Children betray their tendencies in their way of dealing with the breasts that nourish them; nay, I can venture to affirm, that long before they are born they teach their mothers something of their turbulent or quiet tempers.

"Castor gaudet equis, ovo prognatus eodem
Pugnis."

Strike out the false pretensions of phrenology; call it *anthropology;* let it study man the individual in distinction from man the abstraction, the metaphysical or theological lay-figure; and it becomes "the proper study of mankind," one of the noblest and most interesting of pursuits.

The whole physiology of the nervous system, from the simplest manifestation of its power in an insect up to the supreme act of the human intelligence working through the brain, is full of the most difficult yet profoundly interesting questions. The singular relations between electricity and nerve-force, — relations which it has been attempted to interpret as meaning identity, in the face of palpable differences, require still more extended studies. You may be interested by Professor Faraday's statement of his opinion on the matter. "Though I am not satisfied that the nervous fluid is only electricity, still I think that the agent in the

nervous system may be an inorganic force; and if
there be reason for supposing that magnetism is a
higher relation of force than electricity, so it may
well be imagined that the nervous power may be of
a still more exalted character, and yet within the reach
of experiment."

In connection with this statement, it is interesting to
refer to the experiments of Helmholtz on the rapidity
of transmission of the nervous actions. The rate is
given differently in Valentin's report of these experiments and in that found in the Scientific Annual for
1858. One hundred and eighty to three hundred feet
per second is the rate of movement assigned for sensation, but all such results must be very vaguely
approximative. Boxers, fencers, players at the Italian
game of *mora*, " prestidigitators," and all who depend
for their success on rapidity of motion, know what differences there are in the personal equation of movement.

Reflex action, the mechanical sympathy, if I may so
call it, of distant parts ; Instinct, which is crystallized
intelligence,—an absolute law with its invariable planes
and angles introduced into the sphere of consciousness, as *raphides* are enclosed in the living cells of
plants ; Intellect, — the operation of the thinking principle through material organs, with an appreciable
waste of tissue in every act of thought, so that our
clergymen's blood has more phosphates to get rid of
on Monday than on any other day of the week;

Will, — theoretically the absolute determining power, practically limited in different degrees by the varying organization of races and individuals, annulled or perverted by different ill-understood organic changes; — on all these subjects our knowledge is in its infancy, and from the study of some of them the interdict of the Vatican is hardly yet removed.

I must allude to one or two points in the histology and physiology of the organs of sense. The anterior continuation of the retina beyond the *ora serrata* has been a subject of much discussion. If H. Müller and Kölliker can be relied upon, this question is settled by recognizing that a layer of cells, continued from the retina, passes over the surface of the zonula Zinnii, but that no proper nervous element is so prolonged forward.

I observe that Kölliker calls the true nervous elements of the retina " the layer of gray *cerebral* substance." In fact, the ganglionic corpuscles of each eye may be considered as constituting a little *brain*, connected with the masses behind by the *commissure*, commonly called the optic nerve. We are prepared, therefore, to find these two little brains in the most intimate relations with each other, as we find the cerebral hemispheres. We know that they are directly connected by fibres that arch round through the chiasma.

I mention these anatomical facts to introduce a physiological observation of my own, first announced in one of the lectures before the Medical Class, subsequently communicated to the American Academy of Arts and Sciences, and printed in its " Transactions " for February 14, 1860. I refer to the apparent transfer of impressions from one retina to the other, to which I have given the name *reflex vision*. The idea was suggested to me in consequence of certain effects noticed in employing the stereoscope. Professor William B. Rodgers has since called the attention of the American Scientific Association to some facts bearing on the subject, and to a very curious experiment of Leonardo da Vinci's, which enables the observer to look through the palm of his hand (or seem to), as if it had a hole bored through it. As he and others hesitated to accept my explanation, I was not sorry to find recently the following words in the " Observations on Man " of that acute observer and thinker, David Hartley.*

" An impression made on the right eye alone by a single object may‧propagate itself into the left, and there raise up an image almost equal in vividness to itself; and consequently when we see with one eye only, we may, however, have pictures in both eyes." Hartley, in 1784, had anticipated many of the doctrines which have since been systematized into the theory of reflex actions, and with which I have attempted to associate this act of reflex vision. My sixth experiment, however, in the communication referred to,

* Vol. I. p. 207. London, 1801.

appears to me to be a crucial one, proving the correctness of my explanation, and I am not aware that it has been before instituted.

Another point of great interest connected with the physiology of vision, and involved for a long time in great obscurity, is that of the adjustment of the eye to different distances. Dr. Clay Wallace, of New York, who published a very ingenious little book on the eye about twenty years ago, with vignettes reminding one of Bewick, was among the first, if not the first, to describe the *ciliary muscle*, to which the power of adjustment is generally ascribed. It is ascertained, by exact experiment with the *phœnidoscope*, that accommodation depends on change of form of the crystalline lens. Where the crystalline is wanting, as Mr. Ware long ago taught, no power of accommodation remains. The ciliary muscle is generally thought to effect the change of form of the crystalline. The power of accommodation is lost after the application of atropine, in consequence, as is supposed, of the paralysis of this muscle. This, I believe, is the nearest approach to a demonstration we have on this point.

I have only time briefly to refer to Professor Draper's most ingenious theory as to the photographic nature of vision, for an account of which I must refer to his original and interesting Treatise on Physiology.

It were to be wished that the elaborate and very interesting researches of the Marquis Corti, which have revealed such singular complexity of structure in the

cochlea of the ear, had done more to clear up its doubt-
ful physiology; but I am afraid we have nothing but
hypotheses for the special part it plays in the act of
hearing, and that we must say the same respecting the
office of the semicircular canals.

The microscope has achieved some of its greatest
triumphs in teaching us the changes which occur in the
development of the embryo. No more interesting discovery stands recorded in the voluminous literature of
this subject than the one originally announced by Mar-
tin Barry, afterwards discredited, and still later confirmed by Mr. Newport and others; namely, the fact
that the fertilizing filament reaches the interior of the
ovum in various animals; — a striking parallel to the
action of the pollen-tube in the vegetable. But beyond
the mechanical facts all is mystery in the movements
of organization, as profound as in the fall of a stone or
the formation of a crystal.

To the chemist and the microscopist the living body
presents the same difficulties, arising from the fact that
everything is in perpetual change in the organism. The
fibrine of the blood puzzles the one as much as its globules puzzle the other. The difference between the
branches of science which deal with space only, and
those which deal with space and time, is this : we have
no glasses that can magnify time. The figure I here
show you* was photographed from an object (*pleuro-*

---

* From a very interesting paper by Professor O. N. Rood, of Albany,
containing, with other views, the first *microscopic stereograph* I have seen.

sigma angulatum) magnified a thousand diameters, or presenting a million times its natural surface. This other figure of the same object, enlarged from the one just shown, is magnified seven thousand diameters, or forty-nine million times in surface. When we can make the *forty-nine millionth of a second* as long as its integer, physiology and chemistry will approach nearer the completeness of anatomy.

Our reverence becomes more worthy, or, if you will, less unworthy of its Infinite Object in proportion as our intelligence is lifted and expanded to a higher and broader understanding of the Divine methods of action. If Galen called his heathen readers to admire " the power, the wisdom, the providence, the goodness of the Framer of the animal body," — if Mr. Boyle, the student of nature, as Addison and that friend of his who had known him for forty years tell us, never uttered the name of the Supreme Being without making a distinct pause in his speech, in token of his devout recognition of its awful meaning, — surely we, who inherit the accumulated wisdom of nearly two hundred years since the time of the British philosopher, and of almost two thousand since the Greek physician, may well lift our thoughts from the works we study to their great Artificer. These wonderful discoveries which we owe to that mighty little instrument, the telescope of the inner firmament with all its included worlds ; these simple formulæ by which we condense the observations of a generation in a single axiom; these logical analyses by which we fence out the ignorance we cannot reclaim, and fix the limits of our knowledge,—all lead us up to the inspiration of the Almighty, which gives understanding to the world's great teachers. To fear science or knowledge, lest it disturb our old beliefs, is to fear the influx of the Divine wisdom into the souls of our fellow-men; for what is science but the piecemeal *revelation*—uncovering—of the plan of creation, by the agency of those chosen prophets of nature whom God has illuminated from the central light of truth for that single purpose?

The studies which we have glanced at are preliminary in your education to the practical arts which make use of them,—the arts of healing,—surgery and medicine. The more you examine the structure of the organs and the laws of life, the more you will find how resolutely each of the cell-republics which make up the *E pluribus unum* of the body maintains its independence. Guard it, feed it, air it, warm it, exercise or rest it properly, and the working elements will do their best to keep well or to get well. What do we do with ailing vegetables? Dr. Warren, my honored predecessor in this chair, bought a country-place, including half of an old orchard. A few years afterwards I saw the trees on his side of the fence looking in good health, while those on the other side were scraggy and miserable. How do you suppose this change was brought about? By watering them with Fowler's solution? By digging in calomel freely about their roots? Not at all; but by loosening the soil round them, and supplying them with the right kind of food in fitting quantities.

Now a man is not a plant, or, at least, he is a very curious one, for he carries his soil in his stomach, which is a kind of portable flower-pot, and he grows round it, instead of out of it. He has, besides, a singularly complex nutritive apparatus and a nervous system. But recollect the doctrine already enunciated in the language of Virchow, that an animal, like a tree, is a sum of vital unities, of which the cell is the ultimate element. Every healthy cell, whether in a vegetable or an animal, necessarily performs its function properly so long as it is supplied with its proper materials and stimuli. A cell may, it is true, be congenitally defective, in which case disease is, so to speak, its normal state. But if originally sound and subsequently diseased, there has certainly been some excess, deficiency, or wrong quality in the materials or stimuli applied to it. You remove this injurious influence and substitute a normal one; remove the baked coal-ashes, for instance, from the roots of a tree, and replace them with loam; take away the salt meat from the patient's table, and replace it with fresh meat and vegetables, and the cells of the tree or the man return to their duty.

I do not know that we ever apply to a plant any

element which is not a natural constituent of the
vegetable structure, except perhaps externally, for the
accidental purpose of killing parasites. The whole
art of cultivation consists in learning the proper food
and conditions of plants, and supplying them. We
give them water, earths, salts of various kinds such
as they are made of, with a chance to help themselves
to air and light. The farmer would be laughed at
who undertook to manure his fields or his trees with
a salt of lead or of arsenic. These elements are not
constituents of healthy plants. The gardener uses the
waste of the arsenic furnaces to *kill* the weeds in his
walks.

If the law of the animal cell, and of the animal
organism, which is built up of such cells, is like that
of the vegetable, we might expect that we should treat
all morbid conditions of any of the vital unities belonging to an animal in the same way, by increasing,
diminishing, or changing its natural food or stimuli.

" That is an aliment which nourishes ; whatever we
find in the organism, as a constant and integral element, either forming part of its structure, or one of
the conditions of vital processes, that and that only
deserves the name of aliment." * I see no reason,
therefore, why iron, phosphate of lime, sulphur, should
not be considered *food* for man, as much as guano or
poudrette for vegetables. Whether one or another of
them is best in any given case, — whether they shall

* Lewes, Physiology of Common Life, I. 76.

be taken alone or in combination, in large or small quantities, — are separate questions. But they are elements belonging to the body, and even in moderate excess will produce little disturbance. There is no presumption against any of this class of substances, any more than against water or salt, provided they are used in fitting combinations, proportions, and forms.

But when it comes to substances alien to the healthy system, which never belong to it as normal constituents, the case is very different. There is a presumption against putting lead or arsenic into the human body, as against putting them into plants, because they do not belong there, any more than pounded glass, which, it is said, used to be given as a poison. The same thing is true of mercury and silver. What becomes of these alien substances after they get into the system we cannot always tell. But in the case of silver, from the accident of its changing color under the influence of light, we do know what happens. It is thrown out, in part at least, under the epidermis, and there it remains to the patient's dying day. This is a striking illustration of the difficulty which the system finds in dealing with non-assimilable elements, and justifies in some measure the vulgar prejudice against "mineral poisons."

I trust the youngest student on these benches will not commit the childish error of confounding a *presumption against* a particular class of agents with a

condemnation of them. Mercury, for instance, is alien
to the system, and eminently disturbing in its influence.
Yet its efficacy in certain forms of specific disease is
acknowledged by all but the most sceptical theorists.
Even the *esprit moqueur* of Ricord, the Voltaire of
pelvic literature, submits to the time-honored consti-
tutional authority of this great panacea in the class of
cases to which he has devoted his brilliant intelligence.
Still, there is no telling what evils have arisen from the
abuse of this mineral. Dr. Armstrong long ago pointed
out some of them, and they have become matters of
common notoriety. I am pleased, therefore, when I
find so able and experienced a practitioner as Dr. Wil-
liams of this city proving that iritis is best treated without mercury,* and Dr. Vanderpool showing the same
thing to be true for pericarditis.

Whatever elements nature does not introduce into
vegetables, the natural food of all animal life,—directly
of herbivorous, indirectly of carnivorous animals, — are
to be regarded with suspicion. Arsenic-eating may
seem to improve the condition of horses for a time, —
and even of human beings, if Tschudi's stories can be
trusted, — but it soon appears that its alien qualities are
at war with the animal organization. So of copper, antimony, and other non-alimentary simple substances;
every one of them is an intruder in the living system,
as much as a constable would be, quartered in our
household. This does not mean that they may not,

---

* On the Treatment of Iritis without Mercury, Boston, 1856.

any of them, be called in for a special need, as we send
for the constable when we have good reason to think
we have a thief under our roof; but a man's body is
his castle, as well as his house, and the presumption is
that we are to keep our alimentary doors bolted against
these perturbing agents.

Now the feeling is very apt to be just contrary to this.
The habit has been very general with well-taught practitioners, to have recourse to the introduction of these
alien elements into the system on the occasion of any
slight disturbance. The tongue was a little coated,
and mercury must be given; the skin was a little dry,
and the patient must take antimony. It was like sending for the constable and the *posse comitatus* when
there is only a carpet to shake or a refuse-barrel to
empty.* The constitution bears slow poisoning a great
deal better than might be expected ; yet the most intelligent men in the profession have gradually got out of
the habit of prescribing these powerful alien substances
in the old routine way. Mr. Metcalf will tell you how
much more sparingly they are given by our practitioners
at the present time, than when he first inaugurated the
new era of pharmacy among us. Still, the presumption
in favor of poisoning out every spontaneous reaction of
outraged nature is not extinct in those who are trusted

---

* Dr. James Johnson advises persons not ailing to take *five grains of
blue pill* with one or two of aloes twice a week for three or four months in
the year, with half a pint of compound decoction of sarsaparilla every day
for the same period, *to preserve health and prolong life.* Pract. Treatise on
Dis. of Liver, etc., p. 272.

with the lives of their fellow-citizens. " On examining
the file of prescriptions at the hospital, I discovered
that they were rudely written, and indicated a treatment, as they consisted chiefly of *tartar emetic*, ipe-
cacuanha, and epsom salts, hardly favorable to the cure
of the prevailing diarrhœa and dysenteries." * In a
report of a poisoning case now on trial, where we are
told that arsenic enough was found in the stomach to
produce death in twenty-four hours, the patient is said
to have been treated by *arsenic*, phosphorus, bryonia,
aconite, nux vomica, and muriatic acid, — by a practitioner of what school may be imagined.

The traditional idea of always poisoning out disease,
as we smoke out vermin, is now seeking its last refuge
behind the wooden cannon and painted port-holes of
that unblushing system of false scientific pretences
which I do not care to name in a discourse addressed
to an audience devoted to the study of the laws of
nature in the light of the laws of evidence. It is extraordinary to observe that the system which, by its
reducing medicine to a name and a farce, has accus-
tomed all who have sense enough to see through its
thin artifices to the idea that diseases get well without being " cured," should now be the main support
of the tottering poison-cure doctrine. It has unques-
tionably helped to teach wise people that nature heals
most diseases without help from pharmaceutic art, but

---

* United States Sanitary Commission, Document No. 25. Report on a
Regiment near Washington, dated July 9th, 1861.

it continues to persuade fools that art can arrest them all with its specifics.

It is worse than useless to attempt in any way to check the freest expression of opinion as to the efficacy of any or all of the "heroic" means of treatment employed by practitioners of different schools and periods. Medical experience is a great thing, but we must not forget that there is a higher experience, which tries its results in a court of a still larger jurisdiction; that, namely, in which the laws of human belief are summoned to the witness-box, and obliged to testify to the sources of error which beset the medical practitioner. The verdict is as old as the father of medicine, who announces it in the words, "judgment is difficult." Physicians differed so in his time, that some denied that there was any such thing as an art of medicine. One man's best remedies were held as mischievous by another. The art of healing was like soothsaying, so the common people said; the same bird was lucky or unlucky, according as he flew to the right or left.*

The practice of medicine has undergone great changes within the period of my own observation. Venesection, for instance, has so far gone out of fashion, that, as I am told by residents of the New York Bellevue and the Massachusetts General Hospitals, it is almost obsolete in these institutions, at least in medical practice.† The old Brunonian stimulating

* Περὶ Διαίτης 'Οξέων, § IV. v.
† A similar change has taken place also in English surgical practice. Sir W. Napier speaks of "that inveterate use of the lancet, which dis-

treatment has come into vogue again in the practice
of Dr. Todd and his followers. The compounds
of mercury have yielded their place as drugs of all
work, and specifics for that very frequent subjective
complaint, *nescio quid faciam* — to compounds of iodine.* Opium is believed in, and quinine, and "rum,"
using that expressive monosyllable to mean all alco-
holic cordials. If Molière were writing now, instead
of *saignare, purgare,* and the other, he would be more
like to say, *Stimulare, opium dare et potassio-iodizare.*

I have been in relation successively with the English and American evacuant and alterative practice, in
which calomel and antimony figured so largely that,
as you may see in Dr. Jackson's last " Letter," Dr.
Holyoke, a good representative of sterling old-fashioned medical art, counted them with opium and Pe-
ruvian bark as his chief remedies ; with the moderately
expectant practice of Louis ; the blood-letting " *coup
sur coup* " of Bouillaud ; the contra-stimulant method
of Rasori and his followers ; the anti-irritant system
of Broussais, with its leeching and gum-water; I have
heard from our own students of the simple opium
practice of the renowned German teacher, Oppolzer ;
and now I find the medical community brought round
by the revolving cycle of opinion to that same old

graced the surgery of the times," — the early years of this century. Life
and Opinions of Sir Charles James Napier, (London, 1857,) Vol. I. p. 153.

* Sir Astley Cooper has the boldness — or honesty — to speak of medi-
cines which "are given as much to assist the medical man as his patient."
Lectures, (London, 1832,) p. 14.

plan of treatment which John Brown taught in Edinburgh in the last quarter of the last century, and Miner and Tully fiercely advocated among ourselves in the early years of the present. The worthy physicians last mentioned, and their antagonist, Dr. Gallup, used stronger language than we of these degenerate days permit ourselves. "The lancet is a weapon which annually slays more than the sword," says Dr. Tully. "It is probable that, for forty years past, opium and its preparations have done seven times the injury they have rendered benefit, on the great scale of the world," says Dr. Gallup.

What is the meaning of these perpetual changes and conflicts of medical opinion and practice, from an early antiquity to our own time? Simply this: all "methods" of treatment end in disappointment of those extravagant expectations which men are wont to entertain of medical art. The bills of mortality are more obviously affected by drainage, than by this or that method of practice. The insurance companies do not commonly charge a different percentage on the lives of the patients of this or that physician. In the course of a generation, more or less, physicians themselves are liable to get tired of a practice which has so little effect upon the average movement of vital decomposition. Then they are ready for a change, even if it were back again to a method which has already been tried, and found wanting.

Our practitioners, or many of them, have got back to

the ways of old Dr. Samuel Danforth, who, as it is well
known, had strong objections to the use of the lancet.
By and by a new reputation will be made by some dis-
contented practitioner, who, tired of seeing patients die
with their skins full of whiskey and their brains muddy
with opium, returns to a bold antiphlogistic treatment,
and has the luck to see a few patients of note get well
under it. So of the remedies which have gone out of
fashion and been superseded by others. It can hardly
be doubted that they will come into vogue again, more
or less extensively, under the influence of that irre-
sistible demand for change just referred to.

Then will come the usual talk about a change in
the character of disease, which has about as much
meaning as that concerning " old-fashioned snow-
storms." " Epidemic constitutions " of disease mean
something, no doubt; a great deal as applied to mala-
rious affections; but that the whole type of diseases
undergoes such changes that the practice must be
reversed from depleting to stimulating, and *vice versa*,
is much less likely than that methods of treatment go
out of fashion and come in again. If there is any
disease which claims its percentage with reasonable
uniformity, it is phthisis. Yet I remember that the
reverend and venerable Dr. Prince, of Salem, told me
one Commencement day, as I was jogging along towards Cambridge with him, that he recollected the
time when that disease was hardly known; and in
confirmation of his statement mentioned a case in

which it was told as a great event, that somebody
down on " the Cape " had died of " a consumption."
This story does not sound probable to myself, as I
repeat it, yet I assure you it is true, and it shows
how cautiously we must receive all popular stories
of great changes in the habits of disease.*

Is there no progress, then, but do we return to the
same beliefs and practices which our forefathers wore
out and threw away? I trust and believe that there
is a real progress. We may, for instance, return in
a measure to the Brunonian stimulating system, but
it must be in a modified way, for we cannot go back
to the simple Brunonian pathology, since we have
learned too much of diseased action to accept its con-
venient dualism. So of other doctrines, each new
Avatar strips them of some of their old pretensions,
until they take their fitting place at last, if they
have any truth in them, or disappear, if they were
mere phantasms of the imagination.

In the mean time, while medical theories are com-
ing in and going out, there is a set of sensible men
who are never run away with by them, but practise
their art sagaciously and faithfully in much the same
way from generation to generation. From the time
of Hippocrates to that of our own medical patriarch,
there has been an apostolic succession of wise and good
practitioners. If you will look at the first aphorism

---

* See Brit. and For. Med.-Chir. Rev. for Oct., 1860, p. 239. The last two
paragraphs were in type before I had seen the article here referred to.

of the ancient Master, you will see that before all remedies he places the proper conduct of the patient and his attendants, and the fit ordering of all the conditions surrounding him. The class of practitioners I have referred to have always been the most faithful in attending to these points. No doubt they have sometimes prescribed unwisely, in compliance with the prejudices of their time, but they have grown wiser as they have grown older, and learned to trust more in nature and less in their plans of interference. I believe common opinion confirms Sir James Clark's observation to this effect.

The experience of the profession must, I think, run parallel with that of the wisest of its individual members. Each time a plan of treatment or a particular remedy comes up for trial, it is submitted to a sharper scrutiny. When Cullen wrote his Materia Medica, he had seriously to assail the practice of giving *burnt toad*, which was still countenanced by at least one medical authority of note. I have read recently in some medical journal, that an American practitioner, whose name is known to the country, is prescribing the hoof of a horse for epilepsy. It was doubtless suggested by that old fancy of wearing a portion of elk's-hoof hung round the neck or in a ring, for this disease. But it is hard to persuade reasonable people to swallow the abominations of a former period. The evidence which satisfied Fernelius will not serve one of our hospital physicians.

In this way those articles of the Materia Medica which had nothing but loathsomeness to recommend them have been gradually dropped, and are not like to obtain any general favor again with civilized communities. The next culprits to be tried are the poisons. I have never been in the least sceptical as to the utility of some of them, when properly employed. Though I believe that at present, taking the world at large, and leaving out a few powerful agents of such immense value that they rank next to food in importance, the poisons prescribed for disease do more hurt than good, I have no doubt, and never professed to have any, that they do *much good* in prudent and instructed hands. But I am very willing to confess a great jealousy of many agents, and I could almost wish to see the Materia Medica so classed as to call suspicion upon certain ones among them.

Thus the *alien elements*, those which do not properly enter into the composition of any living tissue, are the most to be suspected, — mercury, lead, antimony, silver, and the rest, for the reasons I have before mentioned. Even iodine, which, as it is found in certain plants, seems less remote from the animal tissues, gives unequivocal proofs from time to time that it is hostile to some portions of the glandular system.

There is, of course, less *prima facie* objection to those agents which consist of assimilable elements, such as are found making a part of healthy tissues. These are divisible into three classes, — foods, poisons, and inert, mostly because insoluble, substances. The food of one animal or of one human being is sometimes poison to another, and *vice versa;* inert substances may act mechanically, so as to produce the effect of poisons; but this division holds exactly enough for our purpose.

Strictly speaking, every poison consisting of assimilable elements may be considered as *unwholesome food.* It is rejected by the stomach, or it produces diarrhœa, or it causes vertigo or disturbance of the heart's action, or some other symptom for which the subject of it would consult the physician, if it came on from any other cause than taking it under the name of medicine. Yet portions of this unwholesome food which we call medicine, we have reason to believe, are assimilated; thus, castor-oil appears to be partially digested by infants, so that they require large doses to affect them medicinally. Even that deadliest of poisons, hydrocyanic acid, is probably assimilated, and helps to make living tissue, if it do not kill the patient, for the assimilable elements which it contains, given in the separate forms of *amygdalin* and *emulsin,* produce no disturbance, unless, as in Bernard's experiments, they are suffered to meet in the digestive organs. A medicine consisting of assimilable substances being then simply *unwholesome food,* we understand what is meant by those *cumulative* effects of such remedies often observed, as in the case of digitalis and strychnia. They are precisely similar to the cumulative effects of a salt diet in producing scurvy, or of spurred rye in producing dry gangrene. As the effects of such substances are a violence to the organs, we should exercise the same caution with regard to their use that we would exercise about any other kind of poisonous food, — partridges at certain seasons, for instance. Even where these poisonous kinds of food seem to be useful, we should still regard them with great jealousy. *Digitalis* lowers the pulse in febrile conditions. *Veratrum viride* does the same thing. How do we know that a rapid pulse is not a normal adjustment of nature to the condition it accompanies? Digitalis has gone out of favor; how sure are we that *Veratrum viride* will not be found to do more harm than good in a case of internal inflammation, taking the whole course of the disease into consideration? Think of the change of opinion with regard to the use of opium in *delirium tremens*, (which you remember is sometimes called *delirium vigilans*,) where it seemed so obviously indicated, since the publication of Dr. Ware's admirable essay. I respect the evidence of my contemporaries, but I cannot forget the sayings of the Father of medicine, — *Ars longa, judicium difficile*.

I am not presuming to express an opinion concerning *Veratrum viride*, which was little heard of when I was still practising medicine. I am only appealing to that higher court of experience which sits in judgment on all decisions of the lower medical tribunals, and which requires more than one generation for its final verdict.

Once change the habit of mind so long prevalent among practitioners of medicine; once let it be every-where understood that the presumption is in favor of food, and not of alien substances, of innocuous, and not of unwholesome food, for the sick; that this presumption requires very strong evidence in each particular case to overcome it; but that, when such evidence is afforded, the alien substance or the unwholesome food should be given boldly, in sufficient quantities, in the same spirit as that with which the surgeon lifts his knife against a patient, — that is, with the same reluc-tance and the same determination, — and I think we shall have and hear much less of charlatanism in and out of the profession. The disgrace of medicine has been that colossal system of self-deception, in obedience to which mines have been emptied of their cankering minerals, the vegetable kingdom robbed of all its noxious growths, the entrails of animals taxed for their impurities, the poison-bags of reptiles drained of their venom, and all the inconceivable abominations thus obtained thrust down the throats of human beings suffering from some fault of organization, nourishment, or vital stimulation.

Much as we have gained, we have not yet thor-oughly shaken off the notion that poison is the natural food of disease, as wholesome aliment is the support of health. Cowper's lines, in The Task, show the matter-of-course practice of his time : —

> " He does not scorn it, who has long endured
> A fever's agonies, and *fed on drugs.*"

Dr. Kimball, of Lowell, who has been in the habit of seeing a great deal more of typhoid fever than most practitioners, and whose surgical exploits show him not to be wanting in boldness or enterprise, can tell you whether he finds it necessary to feed his patients on drugs or not. His experience is, I believe, that of the most enlightened and advanced portion of the profession; yet I think that even in typhoid fever, and certainly in many other complaints, the effects of ancient habits and prejudices may still be seen in the practice of some educated physicians.

To you, young men, it belongs to judge all that has gone before you. You come nearer to the great fathers of modern medicine than some of you imagine. Three of my own instructors attended Dr. Rush's Lectures. The illustrious Haller mentions Rush's inaugural thesis * in his Bibliotheca Anatomica ; and this same Haller, brought so close to us, tells us he remembers Ruysch, then an old man, and used to carry letters between him and Boerhaave.† Look through the history of medicine from Boerhaave to

---

* *De Coctione Ciborum in Ventriculo.* Edinb. 1768. — Bibl. Anat. II. 657.

† " Sæpissime bonum senem vidi, sæpe Boerhaavium inter et ipsum literarum vector." Ibid., I. 529.

this present day. You will see at once that medical
doctrine and practice have undergone a long series of
changes. You will see that the doctrine and practice
of our own time must probably change in their turn,
and that, if we can trust at all to the indications of
their course, it will be in the direction of an improved
hygiene and a simplified treatment. Especially will the
old habit of *violating the instincts* of the sick give place
to a judicious *study* of these same instincts. It will be
found that bodily, like mental *insanity*, is best man-
aged, for the most part, by natural soothing agencies.
Two centuries ago there was a prescription for scurvy
containing " *stercoris taurini et anserini par quantitas
trium magnarum nucum*," of the hell-broth containing
which "*quoties-cumque sitit æger, large bibit.*" * When
I have recalled the humane common-sense of Captain
Cook in the matter of preventing this disease; when
I have heard my friend, Mr. Dana, describing the
avidity with which the scurvy-stricken sailors snuffed
up the earthy fragrance of fresh raw potatoes, the *food*
which was to supply the elements wanting to their
spongy tissues, — I have recognized that the perfection
of art is often a return to nature, and seen in this
single instance the germ of innumerable beneficent
future medical reforms.

I cannot help believing that medical *curative* treatment will by and by resolve itself in great measure
into modifications of the food, swallowed and breathed,

---

* Schenck, Observ. Med. Rev., (Lugduni, 1643,) p. 800.

and of the natural stimuli, and that less will be expected from specifics and noxious disturbing agents, either alien or assimilable. The noted mineral-waters containing iron, sulphur, carbonic acid, supply nutritious or stimulating materials to the body as much as phosphate of lime and ammoniacal compounds do to the cereal plants. The effects of a milk and vegetable diet, of gluten bread in diabetes, of cod-liver oil in phthisis, even of such audacious innovations as the water-cure and the grape-cure, are only hints of what will be accomplished when we have learned to discover what organic elements are deficient or in excess in a case of chronic disease, and the best way of correcting the abnormal condition, just as an agriculturist ascertains the wants of his crops and modifies the composition of his soil. In acute febrile diseases we have long ago discovered that far above all drug-medication is the use of mild liquid diet in the period of excitement, and of stimulant and nutritious food in that of exhaustion. Hippocrates himself was as particular about his barley-ptisan as any Florence Nightingale of our time could be.

The generation to which you, who are just entering the profession, belong, will make a vast stride forward, as I believe, in the direction of treatment by natural rather than violent agencies. What is it that makes the reputation of Sydenham, as the chief of English physicians? His prescriptions consisted principally of simples. An aperient or an opiate, a " car-

4

diac " or a tonic, may be commonly found in the midst
of a somewhat fantastic miscellany of garden herbs.
It was not by his pharmaceutic prescriptions that he
gained his great name. It was by daring to order
fresh air for small-pox patients, and *riding on horse-
back* for consumptives, in place of the smothering system, and the noxious and often loathsome rubbish of
the established schools. Of course Sydenham was
much abused by his contemporaries, as he frequently
takes occasion to remind his reader. " I must needs
conclude," he says, " either that I am void of merit,
or that the candid and ingenuous part of mankind, who
are formed with so excellent a temper of mind as to
be no strangers to gratitude, make a very small part
of the whole." *   If in the fearless pursuit of truth
you should find the world as ungracious in the nineteenth century as he found it in the seventeenth, you
may learn a lesson of self-reliance from another utter-
ance of the same illustrious physician : " 'T is none
of my business to inquire what other persons think, but
to establish my own observations ; in order to which, I
ask no favor of the reader but to peruse my writings
with temper." †

The physician has learned a great deal from the
surgeon, who is naturally in advance of him, because
he has a better opportunity of seeing the effects of

---

* Of the Small-Pox and Hysteric Diseases. Epistle to Dr. William
Cole, § 140, Swan's Translation.

† Works, Preface, p. xxi.

his remedies. Let me shorten one of Ambroise Paré's stories for you. There had been a great victory at the pass of Susa, and they were riding into the city. The wounded cried out as the horses trampled them under their hoofs, which caused good Ambroise great pity, and made him wish himself back in Paris. Going into a stable he saw four dead soldiers, and three desperately wounded, placed with their backs against the wall. An old campaigner came up. — " Can these fellows get well? " he said. " No ! " answered the surgeon. Thereupon, the old soldier walked up to them and cut all their throats, sweetly, and without wrath (*doulcement et sans cholere*). Ambroise told him he was a bad man to do such a thing. " I hope to God," he said, " somebody will do as much for me if I ever get into such a scrape " (*accoustré de telle façon*). " I was not much salted in those days " (*bien doux de sel*), says Ambroise, " and little acquainted with the treatment of wounds." However, as he tells us, he proceeded to apply boiling oil of Sambuc (elder) after the approved fashion of the time, — with what torture to the patient may be guessed. At last his precious oil gave out, and he used instead an insignificant mixture of his own contrivance. He could not sleep that night for fear his patients who had not been scalded with the boiling oil would be poisoned by the gunpowder conveyed into their wounds by the balls. To his surprise, he found them much better than the others the next morning, and resolved never again to burn his patients with hot oil for gun-shot wounds.*

This was the beginning, as nearly as we can fix it, of that reform which has introduced plain water-dressings in the place of the farrago of external applications which had been a source of profit to apothecaries and disgrace to art from, and before, the time when Pliny complained of them. A young surgeon who was at Sudley Church, laboring among the wounded of Bull-Run, tells me they had nothing but water for dressing, and he (being also *doux de sel*) was astonished to see how well the wounds did under that simple treatment.

Let me here mention a fact or two which may be of use to some of you who mean to enter the public service. You will, as it seems, have gun-shot wounds almost exclusively to deal with. Three different surgeons, the one just mentioned and two who saw the wounded of Big Bethel, assured me that they found no sabre-cuts or bayonet-wounds. It is the rifle-bullet from a safe distance which pierces the breasts of our soldiers, and not the gallant charge of broad platoons and sweeping squadrons, such as we have been in the habit of considering the chosen mode of warfare of ancient and modern chivalry.†

---

* Le Voyage de Thurin, Œuvres, (Paris, 1579,) p. 1198.

† Sir Charles James Napier had the same experience in Virginia in 1813. "Potomac. We have nasty sort of fighting here, amongst creeks and bushes, and lose men without show." "Yankee never shows himself, he keeps in the thickest wood, fires and runs off." "These five thousand

Another fact parallels the story of the old campaigner, and may teach some of you caution in selecting your assistants. A chaplain told it to two of our officers personally known to myself. He overheard the examination of a man who wished to drive one of the "avalanche" wagons, as they call them. The man was asked if he knew how to deal with wounded men. "O yes," he answered; "if they 're hit *here*," pointing to the abdomen, "knock 'em on the head, — they can't get well."

In art and outside of it you will meet the same barbarisms that Ambroise Paré met with, — for men differ less from century to century than we are apt to suppose; you will encounter the same opposition, if you attack any prevailing opinion, that Sydenham complained of. So far as possible, let not such experiences breed in you a contempt for those who are the subjects of folly or prejudice, or foster any love of dispute for its own sake. Should you become authors, express your opinions freely; defend them rarely. It is not often that an opinion is worth expressing, which cannot take care of itself. Opposition is the best *mordant* to fix the color of your thought in the general belief.

in the open field might be attacked, but behind works it would be throwing away lives." He calls it "an inglorious warfare," — says one of the leaders is "a little deficient in *gumption*," — but "still my opinion is, that if we tuck up our sleeves and lay our ears back we might thrash them; that is, if we caught them out of their trees, so as to slap at them with the bayonet." — Life, etc., Vol. I. p. 218, *et seq.*

It is time to bring these crowded remarks to a close.
The day has been when at the beginning of a course of
Lectures I should have thought it fitting to exhort
you to diligence and entire devotion to your tasks as
students. It is not so now. The young man who has
not heard the clarion-voices of honor and of duty now
sounding throughout the land, will heed no word of
mine. In the camp or the city, in the field or the
hospital, under sheltering roof, or half-protecting can-
vas, or open sky, shedding our own blood or stanching
that of our wounded defenders, students or teachers, —
whatever our calling and our ability, we belong, not to
ourselves, but to our imperilled country, whose danger
is our calamity, whose ruin would be our enslavement,
whose rescue shall be our earthly salvation !

You cannot all follow the armies of your country to
the field. But remember that he who labors for the
general good at home is an ununiformed soldier in the
same holy cause with those who bear arms or minister
at the side of the ambulance and in the camp hospital.
Larrey claimed no precedence of Dupuytren, nor
Guthrie of Sir Astley Cooper. As for the nobleness of
the task for which you are preparing yourselves, I do
not know that I can speak of it more strongly in prose
than I did in the peaceful times before these days of
trial, in the form of verse, and I will so far trespass
on your time and patience as to close this Lecture by
reading you

THE TWO ARMIES.

As Life's unending column pours,
  Two marshalled hosts are seen, —
Two armies on the trampled shores
  That Death flows black between.

One marches to the drum-beat's roll,
  The wide-mouthed clarion's bray,
And bears upon a crimson scroll,
  "Our glory is to slay."

One moves in silence by tne stream,
  With sad, yet watchful eyes,
Calm as the patient planet's gleam
  That walks the clouded skies.

Along its front no sabres shine,
  No blood-red pennons wave ;
Its banner bears the single line,
  "Our duty is to save."

For those no death-bed's lingering shade ;
  At Honor's trumpet-call,
With knitted brow and lifted blade
  In Glory's arms they fall.

For these no clashing falchions bright,
  No stirring battle-cry ;
The bloodless stabber calls by night, —
  Each answers, "Here am I !"

For those the sculptor's laurelled bust,
   The builder's marble piles,
The anthems pealing o'er their dust
   Through long cathedral aisles.

For these the blossom-sprinkled turf
   That floods the lonely graves,
When Spring rolls in her sea-green surf
   In flowery-foaming waves.

Two paths lead upward from below,
   And angels wait above,
Who count each burning life-drop's flow,
   Each falling tear of Love.

Though from the Hero's bleeding breast
   Her pulses Freedom drew,
Though the white lilies in her crest
   Sprang from that scarlet dew, —

While Valor's haughty champions wait
   Till all their scars are shown,
Love walks unchallenged through the gate,
   To sit beside the Throne!

Cambridge : Stereotyped and Printed by Welch, Bigelow, & Co.

www.ingramcontent.com/pod-product-compliance
Lightning Source LLC
Chambersburg PA
CBHW021956190326
41519CB00009B/1292